Marine Metals Manual

A Handbook for Boatmen, Builders and Dealers

by Roger Pretzer

International Marine Publishing Co.
Camden, Maine 04843

Copyright © 1975
by International Marine Publishing Company
International Standard Book Number: 0-87742-055-6
Library of Congress Catalog Card Number: 75-4341
Printed by Evans Printing Company, Concord, New Hampshire.
Bound at New Hampshire Bindery, Concord, New Hampshire.

Second Printing, 1976

Dedication

To Bob Kromer, gifted boat craftsman and respected friend.

Contents

List of Tables   ix
Introduction   xi
Corrosion   1
Aluminum   6
Wrought Copper and Copper Alloys   16
Cast Alloys   25
Magnesium   25
Monel   26
Stainless Steel   27
Steel (Reinforcing Steels for Ferrocement)   35
Titanium (and Titanium Alloys)   43
Zinc   45
Cadmium   47
Wire Rope   47
Chain   51
Purchasing Information   53
Magnets   54
Glossary of Terms   54
Index   61

Tables

Table 1. Galvanic Series   2
Table 2. The four-digit system of aluminum designations for wrought alloys   10
Table 3. Popular marine aluminum alloys   13
Table 4. Cast aluminum alloys   15
Table 5. Wrought copper and copper alloys   20
Table 6. Cast alloys   24
Table 7. Monel wrought alloys   27
Table 8. Stainless steels   28
Table 9. Cast stainless steels   35
Table 10. Ungalvanized steel wire   36
Table 11. Reinforcing bars   38
Table 12. Welded-wire fabric   39
Table 13. Poultry netting   40
Table 14. Woven-wire fabric   41
Table 15. Iron pipe   42
Table 16. Titanium alloys   44
Table 17. Marine wire rope   48
Table 18. Strength of various sizes of chain   52
Table 19. Chain   53

Introduction

There are many types and brands of metals and alloys, as well as vast amounts of technical information describing each one. Rather than make you wade through masses of written technical material, I have presented here a compact listing of metals (and their various aspects) that relates specifically to their use in the contemporary boating field. Included in this listing are definitions, identifications, and strengths of the various metals and alloys; a discussion of corrosion and its prevention; welding information; helpful suggestions; information on wire rope and chain; a section on the reinforcing steels for ferrocement construction; and other valuable information.

This book is intended to help you to recognize quality materials and good workmanship, and to match the best metals with the particular usage you have in mind.

Roger Pretzer

Marine
Metals
Manual

MARINE METALS

If your boat is made of fiberglass, wood, or cement, its critical parts are metal. Think about it: what items on a boat are the most troublesome, costly, and difficult to replace or repair? In what materials is adequate strength hardest to judge without specific facts? What can start to deteriorate the very second you put your boat in the water? Not the fiberglass, wood, or cement — but the little metal parts. The subject of metals may not sound very appealing, but neither does waving your distress flag. Big boats or little boats, it makes no difference — they all depend on metals. The one topic that relates to all metals is corrosion, so we'll begin with that.

CORROSION

Corrosion is not the breakdown of metal by mechanical or physical causes; these are called wear, galling, or erosion. Corrosion is the alteration and decomposition of metals or alloys by persistent electrochemical reactions (involving ion transfer and electron flow) or by direct chemical attack.

Galvanic Corrosion

This type of corrosion causes the most trouble to marine metals. Simply, it works as follows: if two unlike metals are coupled (if they touch each other in an electrolyte, such as salt water), one metal (the less noble) will corrode (oxidize) much faster than it would standing alone; the other metal (the more noble) will not deteriorate.

Table 1. Galvanic Series
(Noble or Cathodic End)

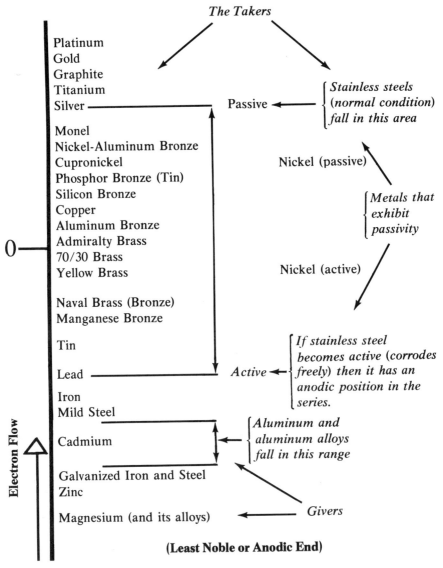

(Least Noble or Anodic End)

[1] The Galvanic Series is often printed in reverse order; either order is acceptable.

[2] Remember that the position of a metal in the Galvanic Series does not

Galvanic Series

Experience and tests have proved that all the known metals can be arranged in what is called the Galvanic Series. Note in the Galvanic Series in Table 1 that platinum and gold are at one end and zinc and magnesium are at the other. For the purposes of this explanation, we will call the metals at the gold end of the list the takers. (Technically, this end is known as the noble or cathodic end.) The metals at the magnesium end of the list we will call the givers (technically, the least noble or anodic end). Note also the various positions of some of the other metals in the series. Using this list, you can generally determine which metal will corrode. For example, if you place zinc next to titanium in salt water, it's good-by zinc; or if you place aluminum next to stainless steel in the presence of an electrolyte, it's good-by aluminum. In general, the farther apart the metals are on the list, the greater the degree of attack.

At this point, some of you may wonder why aluminum is used so extensively in the marine field if it is at the anodic end of the list and, thus, among the most corrodible metals. By itself, aluminum is very noncorrosive (see the section on aluminum). It fails only when other metals are attached directly to it. Stainless steel can safely be attached to aluminum if separators such as plastic washers and sleeves are used to prevent any areas of the two metals from touching. The same holds for any other metals far apart on the list.

Passive-Active

A metal is said to be passive when it remains unchanged for a prolonged period in a corrosive environment (our atmosphere or worse). A metal is said to be active when it corrodes freely.

Passivity

Most metals develop a protective oxide film when exposed to the atmosphere, but some also become much more noble (that is, they move toward the cathodic end of the Galvanic Series). Metals such as iron, nickel, silicon, chromium, and titanium,

---

necessarily reflect its individual resistance to salt water or atmospheric corrosion. Except for iron and steel, most metals have reasonably good or superior corrosion resistance when left unprotected in the environment.

and alloys of these metals, such as stainless steel, are classified as metals that exhibit passivity. (Note the Galvanic Series in regard to stainless steel.) Not subject to this effect are metals such as aluminum, magnesium, and zinc. They, too, develop protective oxide film and can be in a passive or active condition, but they do not exhibit passivity. They more or less retain their same position in the Galvanic Series.

The oxygen concentration required to keep passive the metals that exhibit passivity varies from metal to metal, but there occurs naturally a sufficient amount to accommodate all but iron and iron alloys, which require abnormally high concentrations. Therefore, iron and steel will rust freely unless kept dry and clean or unless coated to prevent attack.

Besides galvanic corrosion, there are several other ways in which metal can deteriorate.

Crevice Corrosion

Crevice corrosion, as explained in the section on stainless steel, affects stainless steel and most other metals, but affects these other metals with lesser degrees of severity. Iron and steel don't need the special features of a crevice to rust (corrode).

To see this type of corrosion, inspect the stainless-steel blade in the safety razor that you shave with. After a week's use, you can generally notice a few small rust spots here and there. If dried soap, hairs, or the razor's casing are left in contact with the blade for a period of time, the oxygen at that particular place will be reduced. If some chemical action or reaction breaks down the stainless steel's protective oxide film under these coverings (crevices), there won't be sufficient oxygen to repair the break and the metal will corrode. Metals subject to crevice corrosion are also subject to pitting.

Pitting

Technically, pitting is a process whereby the deteriorated metal (the pit) becomes an anode and loses metal locally to the much larger passive or cathodic area (the uncorroded area). Pitting can occur both above and below the waterline. A dangerous place for pitting to occur above the waterline is on gas cans that are constantly sprayed with salt water. A localized breakdown of the metal occurs either in shallow, rounded holes or in ever deepening pinholes. The amount of metal lost is usually very small, but pitting on gasoline tanks, cans, or lines can become dangerous because it doesn't take long for a tiny hole to develop. This hole is similar to a radiator leak in a car. Because of the ever present danger of fire and explosion, one should occasionally inspect metal gasoline lines and tanks for this type of corrosion.

Stress or Fatigue Corrosion

Stresses in metal will create avenues and voids within the grain of the metal. These crackled areas tend to corrode more rapidly than unstressed areas. Usually, this type of corrosion is a combination of the action of a corrosive environment and a steady stress in the metal.

Dezincification

This is a sneaky form of corrosion that many people are not aware of. Brasses with a zinc content over 15% are subject to dezincification. Common yellow brass is 65% copper and 35% zinc. Dezincification takes place when brass is placed in a salt-water environment, and intensifies when it is submerged in the water. A sea-water environment will cause the zinc to be carried away as a soluble salt, leaving the metal in an ever weakening condition as this process continues.

There are two types of dezincification. One is the scaling or layer type, and the other is the hole or plug type. The scaling type, the less severe, eats away only about three thousandths of an inch of the brass surface a year. This type is usually found above water on screw and bolt fittings. The hole or plug type is the one to watch out for. The area attacked is relatively small, but the penetration can be as much as one eighth of an inch a year. At this rate, a hole soon will be eaten through all but the most stout brass fitting.

One possible way to spot dezincification is by very close

inspection of the metal to see if any areas are changing from the yellow color of the brass back to the red of copper. Affected areas are softer than the intact metal.

Boats have been known to sink mysteriously when brass through-hull fittings have disintegrated. Make sure you have bronze, plastic, stainless steel, or anything but brass parts below the waterline if you operate your boat in salt water. (See also the section on brass.)

Atmospheric Corrosives

Corrosive agents in the atmosphere contribute to all of the various types of corrosion that take place above the waterline. The most common of these agents are sulphur, chlorine, oxygen, water vapor, and carbon dioxide.

Rusting
Rusting is the corrosion of iron or steel with the formation of visible rust, which consists mainly of watery, scaly ferric oxides. Nonferrous metals, such as aluminum, brass, and zinc, corrode but do not rust.

ALUMINUM

Besides its use in masts, booms, railings, fittings, engines, and engine parts, aluminum is at present the only metal used in large quantities as a structural material in small-boat manufacturing (although the use of steel is expanding because of the growing interest in ferrocement construction). Pure aluminum is not strong enough for structural use, so aluminum alloys have been developed. The alloys of aluminum vary greatly, but it is common to call them simply aluminum. Certain aluminum alloys are considered more suitable for marine use than others, and are usually referred to as marine aluminum. The majority of these alloys are found in the 5000 series (aluminum-magnesium) and the 6000 series (aluminum-magnesium-silicon). Also, some of the alloys in the 7000 series are occasionally used (these are the high-strength aircraft and space-vehicle alloys). The reason these alloys have become known as marine aluminum is because of their high corrosion resistance, low maintenance requirements, and high strength-to-weight ratio.

The alloying elements in aluminum affect that metal's physical and mechanical characteristics in such a way that the alloys divide into two separate groups.

Non-heat-treatable Alloys

The non-heat-treatable alloys (also known as the work-hardening alloys) develop their strength from strain hardening. Manganese and magnesium, singly or together, are the chief alloying elements. Chromium, silicon, zinc, and small amounts of copper are also used in some non-heat-treatable alloys. The non-heat-treatable alloys are in the 1000, 3000, 4000, and 5000 series, and a very few are in the 7000 and 8000 series.

In regard to welding, the non-heat-treatable alloys are generally suitable and usually recommended. Aluminum welding is similar in principle to that of steel, but requires specific knowledge of the alloy to be welded and should be done only by a licensed certified welder. If a question arises that isn't covered in the manuals, a fabricator, aluminum company, or welding supply company should be able to provide the details. Welding methods used are metal-inert-gas arc (MIG) welding and tungsten-inert-gas arc (TIG) welding, the latter being used for thicknesses below one eighth of an inch. In a good welding job, a 100 percent weld efficiency is possible, but you should allow for 10 percent, or more, loss of strength. As in most welding, the surfaces to be welded must be absolutely clean.

Heat-treatable Alloys

The heat-treatable alloys (the 2000, 6000, and 7000 series) obtain additional strength from heat treatment. Basically, these alloys are heated to what is known as their solution-treatment temperature — which lies somewhere between 842° F and 1,022° F (depending on the alloying elements of the particular alloy being made) — and are then quenched. This increases the strength and improves other properties of the alloy through what is known as precipitation hardening. (See *Natural Aging* in the Glossary of Terms.)

Because of the possibility of the severe altering of their strength, heat-treatable alloys are not recommended for welding. There are some exceptions to this, however. In the 6000 series, the alloys 6061, 6062, and 6063 (popular marine-aluminum alloys) can be welded, and often receive postweld heat treatment to improve their strength. The alloys of the 7000 series that are copper-free or contain less than 0.10% copper can be welded. On the other hand, if strength is not an important factor, almost any alloy can be welded. In fact, you could arc-weld in your back yard if high amperages are obtainable from the welding equipment, but the results will most likely be only visual. The best advice is to let a professional do the welding with the proper equipment.

Corrosion

The corrosion resistance of aluminum, which is very good, results from a very thin, hard film of aluminum oxide that forms on the surface of aluminum when it is exposed to the atmosphere. This film thickens with age and is recognized by the dull, grayish look that aluminum assumes after weathering three years or less. Even if the aluminum surface is scratched or sanded, oxidation begins again and a new protective film will build up.

Aluminum is subject to isolated cases of stress corrosion and crevice corrosion, but is most frequently affected by pitting, which occurs when aluminum is exposed to air (atmospheric corrosives) and fresh or salt water. Pitting is most severe in the first few months of exposure of new aluminum to the environment. After the initial exposure, this corrosive action slows down and eventually ceases if the aluminum is allowed to weather.

The two paragraphs above pertain to aluminum when it is isolated. A special corrosion problem with aluminum is contact corrosion (galvanic corrosion). Aluminum is very active (corrodes easily) when it is connected (bolted, screwed, or welded) to most other metals in the presence of an electrolyte. (See the section on galvanic corrosion.) This can be controlled entirely by carefully designing the connections. Rubber or plastic gaskets, bolts dipped in butyl rubber (paint is not recommended), Tufnol washers and sleeves, fiberglass reinforced plastic strips, and special nonhygroscopic tapes (neoprene) are some of the products used as a separative barrier.

Anodizing

Anodizing is the most accepted way of improving the corrosion resistance of aluminum.

Aluminum is the metal usually associated with anodizing because of its special ability to accept a large variety of anodic finishes. In its simplest form, anodizing is the result of a metal being subjected to electrolytic action in order to form a thicker oxide film than would normally be built up through weathering. (Anodizing is not the same as electroplating.) Aluminum is anodized for many reasons, some of the most important of which are added protection from corrosion, resistance to abrasion and wear, and decorative purposes.

If dyes are not used to produce a desired color, the color and final surface texture are determined by the alloying elements in the aluminum. For example, the familiar grayish-gold or yellowish appearance of most anodized sailboat masts is due to the small copper and chromium content of alloy 6061. (6061-T6 is more or less the standard mast material.)

Often, the term "hard-coat anodized" is seen in advertising. This process is one of the two general types of sulfuric-acid anodizing; the other is known as a conventional coating. Hard-coat anodizing has excellent water, wear, and scratch resistance and is much thicker than conventional anodizing.

There are many types of anodizing, and the various aluminum companies have their own particular processes for achieving desired anodized characteristics. If you are interested in knowing the exact advantages of a specific anodized product, you should ask the manufacturer.

Table 2. The Four-Digit System of Aluminum Designations for Wrought Alloys (U.S. System)

Series	Type of Alloy	Suitable usage
1000	99.00% pure aluminum or greater (non-heat-treatable)	low-strength applications
2000	Aluminum-copper (heat-treatable)	aircraft and heavy-engineering construction
3000	Aluminum-manganese (non-heat-treatable)	general usage
4000	Aluminum-silicon (non-heat-treatable)	mostly wire
5000	Aluminum-magnesium (in some alloys, the addition of manganese) (non-heat-treatable)	marine aluminum (good corrosion resistance); structural applications
6000	Aluminum-magnesium-silicon (heat-treatable)	marine aluminum (good for extrusions)
7000	Aluminum-zinc (mostly heat-treatable alloys; a few non-heat-treatable)	where high strength is required (space vehicles and aircraft)
8000	Aluminum-other alloys (beryllium, nickel, tin, titanium)	special purposes
9000	Unused series	

A word about alloys in general is in order at this point. The initial making of an alloy is like cooking stew. Add a little of this and a little of that, until you

Painting

Painting aluminum is not required except for decorative purposes and antifouling protection. Antifouling paints containing mercury, copper, or lead should not be used because of the possibility of galvanic corrosion with aluminum. A three-phase system is often used when bottom painting; it consists of an etch primer, a base paint, and an antifouling paint that is compatible with aluminum. Consult your local marine dealer for your particular choice of primers and paints.

Besides the four-digit system for identifying aluminum and its alloys, there is usually an accompanying temper designation. These designations describe the variety of conditions that determine the strength and other characteristics of an alloy. This is an important part of the identification system because the strengths of individual alloys vary greatly, depending on how and to what degree they are tempered. (See Table 3.)

---

have something that fulfills your requirements. Metallurgists, of course, have kept track of all the properties and all the procedures involved in making a particular alloy, but with so many different people, companies, and possible combinations of materials and procedures, things began to get confusing. So the metallurgists and scientists got together and formed societies and associations to keep abreast of what was going on. They usually adopted or made up a series of numbers or letters, or both, to represent the various alloys. In most cases the numbers or letters mean something; in the rest they are just identification numbers.

Table 2 is a product of what is known as the Aluminum Association. In this table, the numbers do mean something. The first digit indicates the alloy type — aluminum plus almost nothing (for the 1000 series), or aluminum plus the additional major metal or metals (for the 2000, 3000, 4000, etc, series). The last three digits denote one of the many possible combinations of alloy modifications and purity. A detailed explanation of their meanings and ramifications is far beyond the scope of this text, and is of no practical value to the layman. However, if you wish to delve further into the subject of aluminum, visit the engineering section of a library and read *Aluminum* (in three volumes), prepared by engineers, scientists, and metallurgists of the Aluminum Company of America.

Temper Designations for Aluminum and Non-heat-treatable Alloys

F   As fabricated or manufactured (this means you get what you get; there was no control over the temper).

O   Metal is annealed (has low strength, but greatest ductility).

H   The H means that the metal was strain-hardened for strength. It is followed by two or more digits, the first of which tells how it was tempered, and the second, to what degree.

First Digits

H 1   This indicates that the alloy was cold-worked or strain-hardened only.

H 2   This indicates that the alloy was cold-worked and then partially annealed.

H 3   This designation is for aluminum-magnesium alloys only, because these will soften with age if they are not cold-worked and then stabilized by a low-temperature anneal.

Second Digits

Usually, only even numbers are used, but sometimes odd numbers are used to designate intermediate properties. The second digit indicates that the temper is:

H12, H22, H32	Quarter-hard temper.
H14, H24, H34	Half-hard temper.
H16, H26, H36	Three-quarter-hard temper.
H18, H28, H38	Hard temper.
H19, H29, H39	Extra-hard temper.

Usually, the higher the second digit, the greater the strength of that particular alloy.

Third digits are often added to indicate variations of production or control of properties.

Table 3. Popular Marine Aluminum Alloys (Approximate Strengths)

Alloy	Temper	Yield Strength (psi)	Shear Strength (psi)	Available Forms	Usage
5050	H34	24,000	18,000	sheet, plate, drawn tubing	small hulls
	H38	29,000	20,000		
5052	H34	31,000	21,000	sheet, plate, bar, drawn tubing, rod, rivets, forgings	deck houses, small hulls
	H38	37,000	24,000		
5056	H18	59,000	34,000	sheet, rod, rivets	riveting
	H38	50,000	32,000		
5083	H112	23,000	25,000	sheet, plate, extruded shapes, forgings	hull, deck, bulkheads, stanchions
	H321	33,000	28,000		
	H343	41,000	30,000		
5086	H112	19,000	23,000	sheet, plate, drawn tubing, extruded shapes, forgings	welded hull, deck, bulkheads, stanchions
	H32	30,000			
	H34	37,000	28,000		
5454	H112	18,000	23,000	sheet, plate, drawn tubing, extruded shapes	small hulls, hand railings
	H34	35,000	26,000		
	H311	26,000	23,000		
5456	H24	41,000	31,000	sheet, forgings, extruded shapes	welded hulls
6061	T4	21,000	24,000	sheet, plate, extruded shapes, drawn tubing, bar, rod, rivets, forgings	hull, deck, masts, booms, docks, riveting
	T6	40,000	30,000		
	T8	52,000	32,000		
6063	T4	13,000	16,000	drawn tubing, extruded shapes	hand railings, small masts, booms
	T5	21,000	17,000		
	T6	31,000	22,000		
6066	T4	30,000	29,000	extruded shapes, bar, rod, forgings	masts, booms
	T6	52,000	34,000		
6070	T6	52,000	34,000	extruded shapes	masts, booms
6351	T4	27,000	22,000	extruded shapes	hull, deck, stanchions
	T6	43,000	29,000		
7001	T6	91,000		extruded shapes	masts
7075	T6	73,000	48,000	most forms	general-purpose high-strength

Temper Designations for Heat-treatable Alloys

F   As fabricated or manufactured (this means you get what you get; there was no control over the temper).

O   Annealed (the softest temper of wrought products; usually the lowest tensile strength).

T   The T means that the metal was heat-treated to a stable temper other than F or O. The T is also followed by one of more digits, which indicate its particular temper.

T1   Naturally aged to a substantially stable condition.

T2   Annealed (applies to castings only; indicates improved ductility).

T3   Solution heat-treated and then cold-worked, followed by natural aging to a substantially stable condition.

T4   Solution heat-treated and then naturally aged to a substantially stable condition.

T5   Artificially aged only.

T6   Solution heat-treated and then artificially aged. (Sailboat masts usually undergo this process.)

T7   Solution heat-treated and then overaged (improves resistance to corrosion).

T8   Solution heat-treated, then cold-worked, and then artificially aged.

T9   Solution heat-treated, artificially aged, and then cold-worked. (This order of processing improves strength.)

T10   Artificially aged and then cold-worked.

Note:
1. A second or third digit may be added because of a variation in treatment.
2. More often than not, the higher the number after the T, the greater the strength of that particular alloy only (for an example, see the chart "Popular Marine Aluminum Alloys").
3. Remember, this is the designation system used in the United States. It was developed by the Aluminum Association; the British have their own system.
4. See the Glossary of Terms for a definition of *heat treatment.*

Aluminum Alloy Castings

The designation system used by the majority of aluminum producers in the United States is a three-digit system with many additions and subtractions, which makes it confusing. The first digit is the only one with any significance. It places the casting alloy in a specific family, which includes aluminum and the major additional element in that family.

1 to 99	Aluminum-Silicon
100 to 199	Aluminum-Copper
200 to 299	Aluminum-Magnesium
300 to 399	Aluminum-Silicon-Copper
600 to 699	Aluminum-Zinc
700 to 799	Aluminum-Tin
400 to 499	Aluminum-Manganese*
500 to 599	Aluminum-Nickel*

*Note: At present, there are no commercial alloys in these series.

If there has been some variation in the elements from that of the original alloy, a capital letter, usually an A, B, C, or D, is placed in front of the three digits — for example: A132. You would have to refer to the specifications of that particular alloy to find out what the change was. The change is usually the addition of another element.

Table 4. Cast Aluminum Alloys
(Approximate properties of the various casting methods)

Alloy	Sand-Cast		Permanent-mold		Die-Cast	
	Yield	Tensile	Yield	Tensile	Yield	Tensile
380					26,500	45,000
356-T6	24,000	33,000	27,500	38,000		
319-T6	24,000	36,000	27,000	40,000		
A132-T65			43,000	47,000		
D132-T5			27,500	35,000		
113	15,500	24,000				
13					20,000	37,500

An X in front of the series number indicates an experimental or relatively new alloy.

An F is sometimes added to indicate that the alloy is in the as-cast condition. As-cast alloys are those whose mechanical properties are not altered significantly by heat treatment.

Alloys that have been heat-treated (usually to increase strength) have an accompanying temper designation as part of their nomenclature. The procedures they represent are:

T2   Annealed.
T4   Solution heat-treated.
T5   Artificially aged only.
T6   Solution heat-treated and then artificially aged.
T7   Solution heat-treated and then stabilized.

Table 4 lists the most widely used aluminum casting alloys. The strength of a cast product depends on the type of casting method (die, sand, or permanent mold), alloying elements, degree of control over the casting conditions, and temper. Equipment such as cleats, winches, and various fittings are often made from alloy 365 hardened to T6. When you encounter more exotic alloys (there is constant research and refinement in the field), ask the producers to explain their nomenclature and special features.

WROUGHT COPPER AND COPPER ALLOYS

Copper

There are numerous types of copper, most having technical names relating to how they were made, to a special quality, or to the kind of additive they contain. The types most used are electrolytic tough-pitch, oxygen-free, phosphorous-deoxidized, and free-machining (tellurium) copper. (See Table 5 for the properties and uses of wrought copper and copper alloys.)

Copper has good corrosion resistance except in moving sea water; the faster the salt water moves, the more copper corrodes. The above types of copper are fine for applications such as electrical wiring or tubing in on-board fresh-water systems. Places where their use should be avoided if possible are:

WROUGHT COPPER AND COPPER ALLOYS 17

1. In a salt-water tap system. (Salt water is often used for washing when at sea.)
2. In the salt-water-cooling intake tubing of inboard engines.
3. Parts in contact with or containing salt water, such as heat exchangers (condensers) for refrigeration units that operate off an inboard auxiliary engine.
4. Outside heat-exchange tubing for an auxiliary engine.

Cupronickel, which is highly resistant to corrosion due to moving salt water, is recommended for the above four applications. (An alloy of 10% nickel and 90% copper is good.)

The use of copper powder in antifouling paints is diminishing somewhat, due to recent advancements in the field. (Copper gives off an oxide that is toxic to marine microorganisms and that keeps the submerged surface sterilized as long as a sufficient amount of it is leached from the paint.)

Although copper is a very important metal by itself, it is the base metal of brass, bronze, cupronickel and nickel-silver alloys as well. Copper is also added in small percentages to certain types of steel, aluminum, and various other metals as an alloying element. In the marine industry, copper alloys are better known by their names (admiralty brass, silicon bronze, and so forth) than by either their alloy number or their percentages. (Brass is often referred to by the percentages of its components.)

Brass

Add zinc to copper (anywhere from 5% to over 40%) and sometimes add a small percentage of another element such as lead, tin, or manganese and iron (to improve machinability, strength, ductility, or corrosion resistance), and you have the makings of the large variety of brasses. Some of the most common are red brass, yellow brass, cartridge brass, and admiralty brass. Sometimes confusing is the fact that some brasses are called bronzes. Brasses that are called bronzes are:

1. Commercial bronze (90% copper and 10% zinc).
2. Naval brass (often called naval bronze; 60% copper, 39.25% zinc, and 0.75% tin).

3. Manganese bronze (58.5% copper, 39.2% zinc, 1.0% tin, 1.0% iron, and 0.3% manganese)

To keep things clear, remember that the brasses are alloyed with zinc; the major alloying elements of bronzes are tin, aluminum, or silicon.

However, there is a casting alloy called Navy M bronze which is considered a bronze because it is alloyed with tin. Adding a little more to the confusion is the fact that the British sometimes refer to a tin bronze containing 88% copper, 8% tin, and 4% zinc as naval bronze, and to another containing 88% copper, 10% tin, and 2% zinc as admiralty bronze. To keep from being misled when purchasing these particular types of metals, it is best to ask about the alloying elements to determine if you are getting brass or bronze.

Brass with a high zinc content (over 16%) is subject to a type of corrosion known as dezincification. (See the discussion of dezincification in the section on corrosion.) Because of this type of corrosion, brass with a high zinc content is not recommended for fasteners (nuts, bolts, screws, and so forth), through-hull fittings, or any structural application on pleasure boats, especially if the boat is used in salt water. Even some of the manufacturers of boating equipment are not aware of this fact, and produce equipment with brass parts that could easily deteriorate in a salt-water environment.

Three brasses that fall into the high-zinc category and often come under question are admiralty brass (28% zinc), naval brass (39.25% zinc), and manganese bronze (39% zinc). Admiralty brass, if it contains antimony or arsenic as an extra inhibitor to resist dezincification, may be used with some confidence. Naval brass and manganese bronze, although designed to resist dezincification, aren't as reliable. Good advice is not to bother with these types of brass or brass products when stronger and more corrosion-resistant products are available. (See the note below Table 5.)

Brass with a high percentage of zinc is easily identified. As the zinc content of copper increases, the color of the metal changes from a copper-red color (under 5% zinc) to bronze, then through a gold color (10% to 15% zinc), and finally to the light yellow color of brass with a high-zinc content. Cupronickel has an ivory-white or pinkish-white color, and is quite distinguishable from brass.

If you are inspecting a new or used boat, or simply looking for parts, take along a small, fine file, and at an unobvious place on the suspect metal make a small scratch on the brown, tarnished surface. If the underlying metal has a reddish-yellow appearance, it is probably copper or bronze. If the metal is very yellow, further investigation is in order.

Lead added to the various kinds of brasses makes them readily machinable. So if you see leaded this or that brass, you will know that it can be turned with ease on a metal lathe, drilled, sawed, and so forth. Most leaded brasses have a high zinc content of 35% or more.

Most brasses can be welded by oxyacetylene, but not so well by carbon arc.

Brass can be used safely on a boat as a decorative extra (if you don't mind polishing it). One such brass, containing 87.5% copper and 12.5% zinc, closely matches the color of gold. The use of brass for wrought forms and the plating of other metals is popular. Commercial bronze, red brass, and yellow brass, in both wrought and cast forms, can be used for trim. While we are on the subject, other copper alloys that are used for decorative purposes are nickel-silver (12% nickel and 23% zinc) and aluminum-bronze.

Bronze

Bronze differs from brass in that bronze has a high copper content (90.0% or above) and, as previously mentioned, its three main alloying elements are either tin, aluminum, or silicon. There are arguments to this claim because there are many other elements besides, and including, the above three that are added to copper to form the multitude of bronzes. Today, however, the most used, most available, and most accepted are these three types.

Phosphor Bronze

Phosphor bronze originated from the real or original bronzes, which contained copper and tin. These bronzes were

Table 5. **Wrought Copper and Copper Alloys (Approximate properties)**

Popular Name	Yield Strength (psi)	Tensile Strength (psi)	Comments
Aluminum bronze (8% aluminum)	32,000 to 65,000	70,000 to 105,000	Excellent corrosion, wear, and fatigue resistance; has an attractive golden color; available in wire-rope form; excellent for marine hardware; resistant to cavitation erosion.
Aluminum bronze (5% aluminum)	22,000 to 65,000	55,000 to 92,000	
Phosphor bronze (8% tin)	24,000 to 72,000	55,000 to 93,000	Considered a very tough bronze; good elastic qualities; good resistance to cavitation erosion; good fatigue and wear resistance; available in most forms.
Phosphor bronze (5% tin)	20,000 to 75,000	49,000 to 81,000	
Aluminum-silicon bronze (7% aluminum, 2% silicon)	Average 42,000	strength 84,000	Good corrosion and wear resistance; screws, nuts, bolts; good forged turnbuckles.
High-Silicon bronze (3% silicon)	22,000 to 58,000	57,000 to 94,000	General-purpose bronze alloy for nails, screws, nuts, bolts, turnbuckles, thimbles; good corrosion resistance.
Low-Silicon bronze (1.5% silicon)	15,000 to 55,000	40,000 to 70,000	
Commercial bronze (10% zinc)	10,000 to 58,000	37,000 to 67,000	Used for fasteners and trim.
Admiralty brass (28% zinc, 1% tin) with arsenic	18,000 to 70,000	48,000 to 90,000	This is an old-fashioned brass; it has been used for porthole frames and propeller shafts.

[1] The figures in Table 5 represent the comparative strengths of the numerous metals that may be used on pleasure boats. The figures encompass the possible low- and high-strength ranges of tubes, sheets, rods, and strips. The exact strength of a particular wrought stock item or fabricated part depends on the type of material, temper, percentage of the various alloying elements, size, quality control of the manufacturing, and other variables.

[2] When rejecting or accepting a high-zinc alloyed brass part or product,

Popular Name	Yield Strength (psi)	Tensile Strength (psi)	Comments
Cartridge brass (30% zinc)	15,000 to 64,000	47,000 to 78,000	Munitions casements — main use. Has been used in radiator cores and water tanks.
Yellow brass Common (35% zinc)	15,000 to 60,000	47,000 to 74,000	General-purpose brass.
Naval brass (bronze) (39.25% zinc, 0.75% tin)	25,000 to 58,000	55,000 to 75,000	Tensile strength is greater for sections less than ¾" square; has been used for propeller shafts.
Low-leaded brass (0.5% lead, 35% zinc)	15,000 to 60,000	47,000 to 75,000	Good machining and drawing qualities; used for plumbing accessories.
Manganese bronze (brass) (39.2% zinc, 1% tin)	30,000 to 60,000	65,000 to 84,000	High strength and excellent wear resistance, but is subject to dezincification in salt water.
Cupronickel (copper-nickel) (10% nickel)	22,000 to 57,000	44,000 to 60,000	High strength and good ductility; very good corrosion resistance in brackish and moving salt water; available in tubing and some accessories.
Cupronickel (30% nickel)	22,000 to 70,000	55,000 to 77,000	
Electrolytic-Tough-pitch copper and Oxygen-free copper	10,000 to 50,000	32,000 to 55,000	General-purpose coppers; available in most forms.
Deoxidized copper and Free-machining (tellurium) copper	10,000 to 40,000	32,000 to 46,000	

consideration should be given to its size and its location when in service. A screw or bolt that is constantly sprayed or submerged in salt water could cause problems, whereas a large, bulky, manganese-bronze rudder quadrant that is protected below should last several lifetimes.

usually defective because of the presence of oxygen, sulphur, or trapped gases. Oxygen caused the metal to become spongy and weak. Sulphur and trapped gases caused porosity. Oxygen gets into the metal by being absorbed from the air. It can be eliminated by adding to the molten metal something that will combine with the oxygen and then flux off. One such deoxidizing agent that became widely accepted was phosphorus, hence phosphor bronze. Actually, this type of bronze doesn't normally retain any phosphorus, or if it does, it's only a trace. (Aluminum, zinc, manganese, silicon, and antimony help as deoxidizers in the other types of bronzes.) Phosphor bronze has both superior strength and corrosion resistance when compared with brass. It is available in at least five distinct grades and has a good fatigue and wear resistance. Welding is readily accomplished by either TIG or shielded-metal arc methods, but oxyacetylene welding is not recommended.

Silicon Bronze

Silicon bronze is the general-purpose bronze. It has a very high copper content (94% and up) with small amounts of iron, zinc, tin, manganese, and silicon (the highest non-copper element). Two main types are high-silicon and low-silicon bronze. Because of its excellent strength, welding qualities (weldability by just about all methods), and forgeability, you will find that in the marine industry, just about any metal part you can think of can be made from silicon bronze.

Aluminum-silicon Bronze

Aluminum-silicon bronze distinguishes itself from high- and low-silicon bronze by a slight increase in strength and much improved machining and polishing qualities.

Aluminum Bronze

Aluminum bronze is at about the top of the list when comparing the strength and corrosion resistance of the various bronzes. Like the others, it has a high copper content (90% and up), which varies with the different grades. It is also appreciated for its golden color and its elastic properties when hot-worked.

All the above bronzes, especially silicon, are sold under

numerous trade names. If this is the case, the buyer has to rely on the manufacturer's specifications for strengths and other characteristics. All of these bronzes and many more are available in slightly altered cast forms, as well as in wrought forms. Of particular interest to boatmen are propellers, pump bodies, and a vast assortment of hardware. (See Table 6.)

Tobin Bronze

Tobin Bronze is a name often used by naval architects and boat designers when they write specifications for their designs (especially propeller shafts). Anaconda American Brass Company of Waterbury, Connecticut, markets this bronze (Tobin Bronze is a registered trademark). Tobin-Bronze shafting is especially noted for its straightness. Other products made from Tobin Bronze are turnbuckles, various items of marine hardware, piston rings, valve stems, nuts, and bolts.

Cupronickel (Copper-Nickel)

The two most important cupronickel alloys are those with 10% nickel and those with 30% nickel, the latter being superior. They are highly resistant to moving salt water, pitting, and crevice corrosion. They are available in tubing, piping, and fittings. One good advantage they have is that they usually do not require cathodic protection; they are therefore excellent substitutes for copper in such equipment as salt-water desalination apparatus. Recent improvements in the strength of cupronickels, their excellent resistance to corrosion, and the recognition of these qualities by the marine industry should in the near future make them available in more forms.

Chrome Plating

The main purpose of chrome-plating brass and sometimes bronze products is for decorative appearance. Chromium offers some corrosion protection to an underlying metal if it has sufficient oxygen to build and repair the thin oxide film that protects it. (For further explanation, see the sections on either corrosion or stainless steel.) In salt-water boating, chrome-plated products must be protected from constant salt spray if pitting is to be avoided. Their use below the waterline is of little value.

CAST ALLOYS

Like the wrought alloys, the strength of casting depends on many variables. Some important ones are: type of casting (sand, die, permanent mold, and so on), percentage of alloying elements, and quality control of the manufacturing.

MAGNESIUM

Magnesium wins the lightweight contest. If you had 100 pounds of aluminum, you would need only about 65 pounds of magnesium for an equivalent mass. Like most other metals, it is alloyed with other elements (such as aluminum, thorium, and zinc) to improve its strength, machinability, castability, and so forth. The tensile strengths of the various wrought and cast alloys of magnesium are between 23,000 psi and 52,000 psi.

Salt water is one of the places where magnesium is found and from which it is extracted. Unfortunately, magnesium is the metal most subject to galvanic corrosion. If coupled to other metals in salt water, it corrodes severely and returns to its natural state. It occupies the last position at the anodic end of the Galvanic Series (see Table 1). It also pits quite readily by itself. Because of these drawbacks, magnesium is definitely not recommended for any structural use below the waterline.

In the boating world at present, it's most useful as a sacrificial anode to prevent other metals from corroding. However, some attempts have been made to develop coatings to protect it. Magnesium has been used with some success in on-deck equipment because of its light weight. AZ31B and ZE10A are two of the general-purpose alloys available in sheet, plate, and rod form. AZ61A is one popular alloy used in extrusions.

Table 6. Cast Alloys (Approximate properties)

Popular Name	Yield Strength (psi)	Tensile Strength (psi)	Uses
High-strength Yellow Brasses (Manganese Bronze)	30,000 to 70,000	65,000 to 115,000	general high-strength applications, winch handles, propellers, marine hardware
Leaded Yellow Brass	13,000 to 30,000	38,000 to 55,000	boat trim, plumbing, accessories
Tin Bronze	20,000 to 30,000	40,000 to 60,000	pump impellers, bushings, piston rings, bearings, marine hardware
Navy M Bronze (Leaded Tin-Bronze)	16,000 to 20,000	34,000 to 44,000	high-pressure fittings
High-leaded Tin-Bronze	12,000 to 21,000	28,000 to 38,000	bolts, nuts, gears
Silicon Bronze	16,000 to 32,000	36,000 to 54,000	general usage
Aluminum Bronzes	30,000 to 65,000	65,000 to 100,000	marine equipment, struts, shaft logs, propellers
Propeller Bronze (Nickel-Aluminum Bronze)	35,000 to 45,000	85,000 to 100,000	propellers
Manganese-Nickel-Aluminum Bronze	90,000	90,000	propellers
Leaded Nickel Silvers	15,000 to 22,000	30,000 to 45,000	yacht trim, expensive hardware

Magnesium can be welded easily by inert-gas methods. The best joint efficiency is obtained if welding is followed by postweld heat treatment.

MONEL

Monel is a nickel-copper alloy containing approximately two thirds nickel and one third copper. It is considered the best fastener material in the marine industry because of its high strength, high ductility, and excellent resistance to atmospheric and sea-water corrosion. The most common use for monel is in the form of annular-threaded boat nails, screws, nuts, and bolts. It is also an excellent metal for propellers, propeller shafts, and wire rope. (See the section on wire rope for details.)

Monel, like stainless steel and unlike unalloyed copper, is most corrosion-resistant in moving sea water. If the salt water is extremely stagnant for long periods of time, the monel may pit. However, this is a rare problem and need not concern most boatmen. One sign that monel is corroding is the presence of a dull gray-green film on the metal.

Remember that monel is high in the Galvanic Series and can cause other metals such as aluminum to corrode if it is coupled with the other metal in the presence of an electrolyte such as salt water.

Monel can be welded by oxyacetylene, inert-gas arc, and arc welding, but to insure a good weld, special fluxes and special flux-coated electrodes should be used and work should be rapid. Monel can also be brazed or soldered.

Monel is available in sheet, strip, rod, tube, bar, wire, and cast forms. As with most metals, small quantities of additional elements are added to the base (monel alloy 400) to improve such things as machining qualities, strength, and castability. Table 7 lists the types of monel of particular interest to boatmen.

Cast Monel Alloys

The quality of a cast product depends largely on the molten metal's ability to flow easily and fill all the voids in the mold. The

Table 7. Monel Wrought Alloys

Alloy	Yield Strength (psi)	Tensile Strength (psi)	Comments
Monel-400	25,000 to 130,000	70,000 to 140,000	Higher strengths available in hardest drawn wire. Slightly magnetic. Available in most forms. General purpose base alloy.
Monel-R-405 (strengths given here are for bar and rod)	25,000 to 110,000	70,000 to 135,000	A small amount of sulfur has been added to make this the free-machining alloy.
Monel-K-500 (with small amounts of aluminum and titanium)	40,000 to 175,000	90,000 to 200,000 (or higher)	Nonmagnetic; has higher strengths because it can be age-hardened; available in most forms.

addition of more silicon to monel improves its fluidity. The strength of a casting can also be controlled with he aid of hardening techniques and additions of silicon. Yield strengths of 115,000 psi and tensile strengths of 145,000 psi are obtainable in some cast monel products. Monel alloy 410 (with 0.5% to 1.6% silicon), monel alloy 506 (2.5% to 3.0% silicon), and, the strongest of the three, monel alloy 505 (3.5% to 4.5% silicon) are used for marine casting purposes.

STAINLESS STEEL

Stainless steels divide into three general classes: martensitic, ferritic, and, the group most suitable for marine applications, austenitic. All are iron alloys in which chromium is the major alloy, varying from about 11.5% to 26%. Other important elements contained in all three classes (in very small percentages) are carbon, phosphorus, sulfur, manganese, and silicon.

Table 8. Stainless Steels (Approximate Properties of Standard Grades)

AISI Type	Yield Strength (psi)	Tensile Strength (psi)	Comments	Uses
301 17% Cr 7% Ni	40,000 to 140,000	110,000 to 185,000	high-strength and light-weight applications	tangs, straps, wire
302 18% Cr 8% Ni	37,000 to 150,000 (up to 350,000 for cold-drawn wire)	90,000 to 180,000	general-purpose	nuts, bolts, fishing tackle, rigging wire
303 18% Cr 8% Ni	35,000 to 145,000	90,000 to 180,000	good where extensive machining is required	custom-made parts and fittings, screws, nuts, bolts
304 18.5% Cr 8.5% Ni	35,000 to 150,000	85,000 to 180,000	general-purpose; good welding; good corrosion resistance	railings, rigging wire, structural applications where welding is required
305 18% Cr 11.5% Ni	37,000 to 95,000 (or higher)	85,000 to 150,000	good corrosion resistance and good cold-forming ability	tangs, straps, high-strength nuts, substitute for type 302
316 17% Cr 12% Ni 2.5% Mo	35,000 to 125,000 (up to 300,000 for some cold-drawn wire)	85,000 to 150,000	excellent corrosion resistance, especially underwater	propeller shafts, fittings, bolts, nuts, wire
317 19% Cr 14% Ni 3.5% Mo	40,000 to 95,000 (or higher)	90,000 to 120,000	excellent corrosion resistance; becomes slightly magnetic if cold-worked	propeller shafts, substitute for type 316

The stainless steels listed in Table 8 are the ones most used or best suited for pleasure boats. Low figures are for the metal in the annealed condition (the condition that is less subject to corrosion). The steels acquire greater strength from tempering of the wrought stock or from being worked during fabricating processes. Only the manufacturer can certify the exact strength of his merchandise. (In this case, tempering denotes both cold-worked tempering and precipitation hardening.)

AISI Type	Yield Strength (psi)	Tensile Strength (psi)	Comments	Uses
321 18.5% Cr 10% Ni 4% Ti	35,000 to 125,000	87,000 to 150,000	superior for welded equipment exposed to a corrosive environment	welded fabricated parts
347 18.5% Cr 10% Ni	35,000 to 125,000	92,000 to 150,000	excellent welded corrosion resistance	very heavy welded fabrications
201	55,000 to 140,000	115,000 to 185,000	nonmagnetic	substitute for type 301
202	55,000 to 75,000	105,000 to 125,000 (or higher)	nonmagnetic	substitute for type 302
17-4PH (type 630)	110,000 to 185,000	150,000 to 200,000	magnetic; also a popular casting alloy	tangs, straps, blocks, winch parts, general usage
17-7PH (tpe 631)	40,000 to 250,000	130,000 to 265,000	magnetic in the high-strength condition	open for experimentation in the marine industry
17-10P	38,000 to 98,000	89,000 to 144,000	nonmagnetic	open for experimentation in the marine industry
AM-355 (type 634)	55,000 to 210,000	160,000 to 230,000	magnetic in the high-strength condition	shafting

Martensitic stainless steels have the distinction of being the only stainless steels that can be hardened by heat treatment. This, along with other factors, contributes to their ability to hold a good edge, which makes them ideal for use in cutlery. They are also used in valves, ball bearings, and nozzles. A characteristic of martensitic stainless steels that is sometimes thought objectionable is that they are magnetic.

Ferritic stainless steels are not hardenable by heat treatment, and their use is generally limited to trim or decorative applications (automobile trim, for instance) because of the ease with which they can be buffed to a shiny finish. Ferritic stainless steels are also magnetic.

18-8 Grades

Austenitic stainless steels are characterized by the addition of a second major alloying element — nickel. One familiar system of identifying austenitic stainless steels is to refer to them as 18-8 grades (18% chromium and 8% nickel). There are some variations in these percentages, but the group, both as a whole and individually, is usually referred to as 18-8.

300 Series

The identification system preferred by the boating industry for the wrought types of stainless steel is that of the American Iron and Steel Institute (AISI). The part that covers most of the austenitic stainless steels is known as the 300 series. The 200, 400, and 600 series contain just about all the rest of the stainless-steel alloys. The stainless steels in the 300 series are nonmagnetic, can be strengthened by cold-working, and have excellent resistance to corrosion — which makes them advantageous for marine use.

The most widely used stainless steels in the 300 series are types 302, 303, and 304. The majority of bolts, nuts, rigging cables, and chain plates are made from these types. A higher carbon content in type 302 enables that type to resist fatigue better than type 304. Then again, the lower carbon content of type 304 makes it a better material for welding. Type 316 isn't as strong as type 302, but a slight increase in size of the particular part in question usually compensates for this. Example: Where you would use a ¼" bolt made of 302, you should consider a 5/16" bolt made of type 316. It should be pointed out that the numbers used to identify the various stainless steels have no more logical meaning than the channels of your television set, except that most of the austenitic stainless steels have been placed in the 300 series. Table 8 lists the various properties and strengths that identify the grades of stainless steel.

Corrosion

The chromium content of stainless steel is the major contributor to its corrosion resistance. The addition of nickel and molybdenum also helps corrosion resistance, but higher carbon percentages tend to have a slight negative effect.

The reason why stainless steel is so corrosion-resistant is quite fascinating. Stainless steel is one of the metals that exhibit what is known as passivity. In other words, it becomes more noble (more like gold) when certain conditions exist. Fortunately, our atmosphere supplies these conditions. However, some manufacturers of stainless-steel products fortify the passive state of the metal by artificial means, and term the resulting product passivated. ("Our stainless products are passivated.") A sufficient concentration of oxygen is present in our atmosphere to form or repair a tough, transparent chromium-oxide film on the surface of stainless steel that renders the metal noncorrodible. There is even enough oxygen in most salt water to keep most stainless steel in its passive state. So then, how can stainless steel corrode?

Crevice Corrosion

There are several ways that corrosion can take place. One possibility is when the oxygen volume at a particular place is indirectly cut down. For example, this can happen between a bolt head and a washer that are under salt water. This isolated area of stainless steel may not have enough oxygen to rebuild the chromium-oxide film if the film is scratched or worn away due to minor movements or vibrations. If this happens, the stainless steel passes into its active state and corrodes freely. (It can corrode as fast as a piece of steel exposed to rain.) This type of corrosion is often referred to as crevice corrosion or a form of concentrated cell corrosion. The possibility of it taking place is reduced if the boat or salt water is in motion. Brackish or stagnant salt water, because of its lack of oxygen, will enhance attack.

Pitting

The most common way stainless steel corrodes is by pitting. The chloride ions in salt water break down very tiny areas of the protective film on the surface of stainless steels. In time, these random active areas become slowly deeper as they continue to deteriorate. Pits are at least as deep as the diameter of the pit opening; more often, they are far deeper, sometimes like a pinhole.

All stainless steels are susceptible to crevice corrosion and pitting, but some more so than others. Tests have shown that the severity of pitting of type 316 diminishes after its initial submersion in salt water, compared to a steady increase for type 302. (Type 316 has a small amount of molybdenum added to it to improve its corrosion resistance.) This quality makes type 316 the best stainless steel for underwater applications, such as propeller shafts.

Staining

Another problem encountered with stainless steel in ocean environments is staining. This is recognizable as a light brown film that looks like rust and forms on the surface of the metal. If the metal is attached to a painted area of a boat, it may bleed and leave streaks down it. Restoring the metal to its shiny appearance is easily accomplished by applying a rust solvent.

Galvanic Corrosion

Stainless steels in a passive state (normal condition) are cathodic and will make other metals such as aluminum (which are anodic) corrode away if they are in or around salt water. For this reason, a nonmetallic separator is often used in connecting stainless steel with an anodic metal.

Welding

Fortunately for boatmen, the austenitic type of stainless steel (200 and 300 series) is the most weldable. Besides strength, the most important requirement in welding stainless steel for marine use is keeping the weld from losing its stainless qualities. Sensitization (a problem that arises when the austenitic grades are heated to between 800° F and 1,500° F) and the contamination of the welded area by oxygen are the two main contributors to eventual corrosion headaches. Both of these obstacles can be avoided by a thorough understanding of what causes them, along with using the proper welding equipment and techniques.

In factory production lines, many problems are overcome by the use of resistance welding. In this type of welding, no filler metal is required; the hot joints are merely pressed together. Arc-welding techniques such as TIG arc, MIG arc, and shielded-metal arc are suitable for welding stainless steel in or out of the factory. (For explanations of these welding processes, see the Glossary of Terms.)

Filler metals in the form of wires or welding rods are quite specialized, and a particular type is prescribed for each grade of stainless steel. The weld should be ground down in order to erase or smooth the cracks and lumps; this also helps render the weld more corrosion-resistant.

Other Important Stainless Steels for Marine Use

Types 201 and 202 — which have manganese in place of part of the nickel, and nitrogen as a stabilizer — are austenitic stainless steels that can replace types 301 and 302. These steels were developed to save nickel.

Precipitation-hardened stainless steels are receiving some acceptance from yachtsmen. These steels are recognizable by their trade designation or their AISI numbers. The most popular is 17-4PH (type 630), which is used in blocks and tackles, winches, and in mounting straps and tangs. This type can be considered a high-strength stainless steel. Others worth investigation are 17-7PH (type 631), AM-355 (type 634), and

17-10P. It's surprising that the marine industry doesn't use more precipitation-hardened stainless steels. Most of them have higher strengths and better corrosion resistance than their previously discussed brothers. However, some are magnetic. There are several manufacturers that produce precipitation-hardened stainless steels, but the majority of them are staying with the more familiar 300 grades. (For further details, see Table 8.)

Cast Stainless Steels

Table 9 lists important casting alloys and their wrought counterparts. The alloying-element specifications for cast stainless steel are those of the Alloy Casting Institute (ACI). They are not exactly the same as the AISI specifications for wrought stainless steel, but there is enough of a similarity so that one can assess the relative quality of the wrought and cast stainless steels.

In the example that accompanies Table 9, the letter "C" stands for corrosion-resistant casting. This refers to castings designed to resist corrosive attack at temperatures less than 1200° F. The letter "H" as the first prefix in an ACI casting type stands for heat-resisting alloy. The "H" types are not often recommended for marine use.

The second letter indicates the approximate percentages of chromium and nickel used. As you go down the alphabet, the amounts increase. For example, G indicates a higher chromium-nickel content than F: F represents about 18% to 19% chromium and 8% to 9% nickel, and G, about 19% to 21% chromium and 10% to 12% nickel.

The number stands for the maximum carbon content in hundredths of a percent.

If a special element or elements are added, they are indicated by an extra letter:

    C — Columbium
    M — Molybdenum
    F — Free-machining (easily worked by machines); this is achieved by adding very small amounts of selenium, phosphorus, and molybdenum.

Table 9. Cast Stainless Steels
(Approximate properties)

Wrought AISI type	Cast ACI type	Yield strength (psi)	Tensile strength (psi)
302	CF-20	35,000	77,000
303	CF-16F	40,000	76,000
304	CF-8	37,000	76,000
316	CF-8M	42,000	80,000
316	CF-12M	42,000	80,000
317	CG-8M	40,000	82,000
347	CF-8C	38,000	77,000

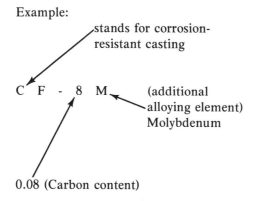

Example: C (stands for corrosion-resistant casting) F - 8 M (additional alloying element) Molybdenum; 0.08 (Carbon content)

STEEL (Reinforcing Steels for ferrocement)

Besides being used in engines and some hardware items, steel is becoming an ever more popular structural material in pleasure boats. Although occasionally used in sheet form for hull plating, it has most recently become popular as structural reinforcement in ferrocement boat construction.

There is a large variety of reinforcing steels from which to choose. This doesn't mean that all are practical and economical, but since there is still considerable experimentation and

Table 10. Ungalvanized Steel Wire (Approximate properties)

Gauge (or fraction of an inch)	Diameter in inches	Pounds per 100 feet	Number of feet per pound	Cross-sectional area in square inches	Breaking strength in pounds
3/8"	.375	37.5	2.66	.1104	10,900
000	.3625	35.05	2.85	.1032	10,200
00	.3310	29.22	3.42	.0860	8,600
0	.3065	25.06	3.99	.0738	7,400
1	.2830	21.36	4.68	.0629	6,300
2	.2625	18.38	5.44	.0541	5,400
1/4"	.2500	16.67	6.0	.0491	4,900
3	.2437	15.84	6.31	.0466	4,650
4	.2253	13.54	7.40	.0398	3,950
5	.2070	11.43	8.75	.0336	3,350
6	.1920	9.83	10.17	.0289	2,850
3/16"	.1875	9.40	10.65	.0276	2,700
7	.1770	8.36	12.0	.0246	2,450
8	.1620	7.0	14.3	.0206	2,050
9	.1483	5.87	17.0	.0173	1,700
10	.1350	4.86	20.6	.0145	1.400

development taking place, especially in the direction of thinner hulls (a half inch or less for smaller boats), all can and should be considered.

When designating a type of reinforcing steel, be aware that the same number may refer to different sizes, depending on what product it describes. For example, note that the diameter of a 3-gauge steel wire (.2437 inches) is substantially smaller than a number 3 reinforcing bar (0.375 inches). Also, (though not listed here) except for the fractions of an inch (3/8, 1/4, 3/16, etc.), the thicknesses of sheet metals are not the same as diameters of steel wires with similar gauge numbers.

Familiarization with Table 10 should help you decipher and understand the subsequent tables.

STEEL 37

Gauge (or fraction of an inch)	Diameter in inches	Pounds per 100 feet	Number of feet per pound	Cross-sectional area in square inches	Breaking strength in pounds
1/8"	.1250	4.17	24.0	.0123	1,200
11	.1205	3.87	25.8	.0114	1,125
12	.1055	2.97	33.7	.0087	860
13	.0915	2.23	44.8	.0066	660
14	.800	1.71	58.6	.0050	500
15	.720	1.38	72.3	.0041	400
16(1/16")	.0625	1.04	96.0	.0031	310
17	.0540	0.778	128.6	.0023	225
18	.0475	0.602	166.2	.0018	170
19	.0410	0.448	223.0	.00132	130
20	.0348	0.323	309.6	.00095	95
21	.0317	0.268	373.1	.00079	80
22	.0286	0.218	458.4	.00064	60
23	.0258	0.175	563.3	.00052	49
24	.0230	0.141	708.7	.000415	42
25	.0204	0.111	902.0	.000326	31

[1]The steel wire gauges (and corresponding diameters) listed here are the standard gauges used by most U.S. manufacturers. (There are many different sets of diameters with the same gauge numbers as the ones in this table. They may vary because of the type of metal used to make the wire, or because of the country of origin.)

[2]The weights, diameters, and other figures in this table are for plain uncoated steel wire. Galvanized wire is available but its weights and properties are different from those listed in this table.

[3]In some cases, galvanizing may weaken the wire somewhat. If the wire hasn't been fully annealed, the heat from the galvanizing process may tend to do so. (Annealing relieves stresses caused by the drawing of the wire, with a subsequent reduction in the tensile strength.)

[4]Half gauges (11½, 12½, 14½, and so on) are available, though they aren't usually stock items.

Table 11. Reinforcing Bars ("Re-Bars")

Type	Number	Diameter in inches		Weight in pounds per foot	Cross-sectional area in square inches
Plain round	2	.250	1/4"	0.17	.05
Deformed	3	.375	3/8"	0.38	.11
	4	.500	1/2"	0.67	.20
	5	.625	5/8"	1.04	.38
	6	.725	3/4"	1.50	.44

[1] The bar number is the number of eighth-inch segments in the diameter of the reinforced bar.

[2] Three or more grades of reinforcing bars are generally available. Grades are distinguished by minimum yield strengths. Popular ones are 40,000 psi, 60,000 psi, and 75,000 psi, known respectively as Grade 40, Grade 60, and Grade 75.

[3] The bars that are higher in yield and strength are quite springy and hold their shape well. They are best suited for the longer runs that require only moderate bending.

[4] The bars that are lower in yield strength are bent much easier and are better suited for the sharp reverse bends required in forming truss frames and webs. (This type of shape has been nicknamed "wiggle rod.")

[5] Bars come in 20- and 40-foot lengths; lengths of 60 feet and longer are available, but they present transportation problems.

[6] Deformed bars (bars that have projecting ribs or ridges that improve the bend between the steel and cement) are usually available only in size 3 and up.

Styles of Welded-Wire Fabric
(also referred to as reinforcing mesh)

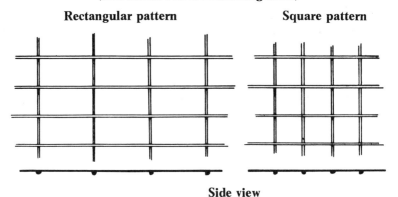

Rectangular pattern Square pattern

Side view

Table 12. Welded-Wire Fabric

Style Designation	Weight per 100 square feet in pounds	Style Designation	Weight per 100 square feet in pounds
1 x 1 - 16/16	26	3 x 3 - 12/12	25
1 x 1 - 14/14	42	3 x 3 - 10/10	41
1 x 1 - 12/12	74	3 x 3 - 8/8	58
1 x 2 - 16/16	20	4 x 4 - 14/14	11
1 x 2 - 14/14	32	4 x 4 - 12/12	19
2 x 2 - 16/16	13	4 x 4 - 10/10	31
2 x 2 - 14/14	21	4 x 4 - 8/8	44
2 x 2 - 12/12	37	4 x 4 - 6/6	62
2 x 2 - 10/10	60	4 x 8 - 12/12	14
2 x 4 - 14/14	16	6 x 6 - 10/10	21
2 x 4 - 12/12	28	6 x 6 - 8/8	30
2 x 4 - 11/11	36	6 x 6 - 6/6	42
3 x 3 - 16/16	10	6 x 6 - 4/4	58
3 x 3 - 14/14	14	6 x 6 - 2/2	78

[1] In the style designation columns, the first two numbers (1 x 1, 2 x 2, and so on) represent the spacing in inches between the wires, first in the longitudinal direction, then in the transverse. The second two numbers (16/16, 14/14, and so on) represent the wire gauge.

[2] This type of reinforcing for ferrocement consists of cold-drawn wires (see Table 10 for other properties) and is held together at all intersections by welds. Welded-wire fabric can replace reinforcing rods that have been used as a base or center core to which chicken wire or woven wire is attached. The completed wire structure (the hull before cement is applied) has been nicknamed the "armature."

[3] Welded-wire fabric made with larger wire comes in sheets. The more flexible type, made from smaller-diameter wire, is available in rolls with varying widths and lengths. As with Table 10, this table lists ungalvanized material, though galvanized forms are available.

[4] Welded wire fabric is available in either square or rectangular patterns. The square pattern has been the most widely used because of the assumption that it provides an equal distribution of strength. More recent thinking, however, has suggested that equal distribution of strength in ferrocement boat hulls may not always be advantageous. Therefore, a variety of rectangular patterns has also been listed.

Table 13. Poultry Netting
(also known as chicken wire or hexagonal mesh)

Mesh Opening	Identified as	Various steel wire gauges that mesh is (or could be) available in	Weight in pounds of 100 square ft. of 20-gauge netting
1/2	1/2-inch mesh	19, 20, 21, or 22	15.3
1	1-inch mesh	19, 20, 21, or 22	10.2
2	2-inch mesh	19, 20, 21, or 22	5.0

[1] Poultry netting (in rolls) is available in a variety of widths up to 6 feet, and is usually 150 feet long.

[2] It generally comes galvanized, but is obtainable ungalvanized. Galvanizing is usually done after the netting is formed. This stiffens the material somewhat because the zinc tends to hold the twisted joints tightly.

[3] The most common size is the one-inch mesh. The half-inch mesh preferred by some builders is harder to obtain because not many suppliers stock it.

[4] Chicken wire was the favorite of the early ferrocement boatbuilders because it adapted well to the curves of hulls. Eight layers (four layers on each side of the central core of either pipe frames or reinforcing bars) has been the usual standard.

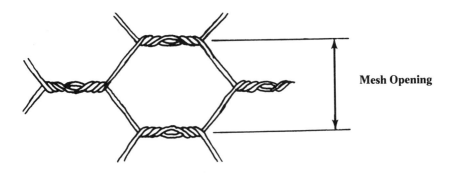

Table 14. Woven-Wire Fabric (also known as wire cloth or mesh)

Number of wires per linear inch	Commonly known as (style designation)	Also known as	Various steel-wire gauges that woven fabric is (or could be) available in
1	1 x 1		3, 5, 8, 10, 12, or 14
1 wire every 3/4"	3/4 x 3/4		4, 6, 8, 10, 12, 14, or 16
1 wire every 5/8"	5/8 x 5/8		5, 7, 10, 11, 12, 14, or 16
2	2 x 2	1/2 x 1/2	6, 8, 11, 14, 16, 18, or 19
3	3 x 3		9, 11, 13, 16, 17, 19, or 21
4	4 x 4	1/4 x 1/4	11, 12, 14, 16, 18, 20, or 23
5	5 x 5		12, 14, 16, 17, 18, 21, or 24
6	6 x 6		14, 15, 16, 18, 20, 22, or 25
7	7 x 7		15, 16, 18, 20, 21, 22, or 26
8	8 x 8		16, 17, 18, 20, 22, 23, or 27

[1] The standard types of wire fabric (in rolls) are available in widths up to 6 feet.

[2] Woven-wire fabric is one of the most diversified types of reinforcing available, though until now it has been overlooked somewhat by ferrocement boatbuilders. Woven-wire fabric is not welded, but is held together by the weave and the galvanizing; it is obtainable in the uncoated condition, however.

[3] Besides the square mesh listed here, a diamond-shaped mesh is manufactured.

[4] Recently, boatbuilders tend to prefer the square pattern of woven-wire fabric over poultry netting for the outer wires of the armature. A much larger selection of wire (especially finer grades) and the obtainability of a more even distribution of wire are two reasons for this.

(*Wire fabric with openings under 1/4 inch is referred to as metal cloth*)

Square pattern

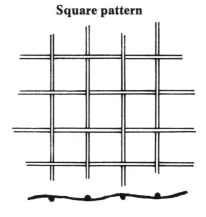

Table 15. Iron Pipe

Nominal pipe size (opening)	Outside diameter	Wall Thickness		
		Light	Standard	Extra strong
1/8"	0.405	.049	.068	.095
1/4"	0.540	.065	.088	.119
3/8"	0.675	.065	.091	.126
1/2"	0.840	.083	.109	.147
3/4"	1.050	.083	.113	.154

[1]Black wrought (as opposed to cast) iron pipe is the original stand-by for framing by many former ferrocement boatbuilders. The problems with iron pipe are internal. The pipes have to be filled with grout in order to prevent untraceable leaks, to muffle the noise from mysterious objects rolling around in the boat, and to guard against rusting from the inside out. If care isn't taken to fair the hull properly during the cementing, there will be inside and outside ridges at the pipe-frame stations that will leave the hull with a rumpled look.

[2]The outside diameter is the actual size of the pipe.

Expanded Metal

This type of reinforcing (generally used by plasterers) comes in a variety of sizes, patterns, and styles, of which diamond mesh and expanded rip lath are generally the most popular. It's formed from thin steel sheets that are slit and then pulled apart (expanded) to form the netlike patterns. Some builders have found that this material doesn't conform to curves very well.

Nature of Ferrocement

Ferrocement is akin in many respects to fiberglass. The steel reinforcing serves the same function as the fiberglass cloth, and the cement plays the same part as the plastic resins. The coefficient of expansion is about the same for steel as for cement so expansion and contraction due to temperature changes should not cause cracking. Of course, bending can cause the cement to crack. Placing many fine wires as near to the surface as possible helps alleviate this problem. However, too many, too fine wires near the surface greatly inhibit thorough cement penetration. I hope that this overall presentation of the various possibilities of reinforcing steel for ferrocement will generate more experimentation.

It is useful to list and talk about all the various steel reinforcing products that could be used for ferrocement construction, but the real problem is finding a place to purchase them, especially the fancy ones. Don't give up looking if a local building-supply outlet doesn't handle them. Call the large wire or metal companies; they will deal with you whether your order is large or small. One of the best ways to become acquainted with suppliers and manufacturers is to visit your local library (see the section on purchasing information).

TITANIUM (and Titanium Alloys)

Titanium is a unique metal. It has a very high strength-to-weight ratio (it's as strong as steel, but weighs only half as much), and is more resistant than stainless steel or silver to galvanic corrosion in salt water. This good corrosion resistance is due to a self-repairing oxide film that forms on the surface of the metal. Tests in salt water have shown that it would take one thousand years to corrode one one-thousandth of an inch. It increases the corrosion of most other metals when in contact with them, but is not affected itself (see the section on galvanic corrosion).

The melting point of titanium is 3,137° F, compared with 2,802° F for iron. This caused quite a problem in its early stages of development because it could melt the containers it was made in. Now, processing is performed in an electric furnace in a special container that has a water-cooled lining and an argon atmosphere to prevent contamination by oxygen. Titanium becomes very brittle when melted in the open air because of its ability to absorb large amounts of oxygen and nitrogen.

Titanium is generally referred to as just "titanium," but as a pure metal it has a lower ultimate strength and ductility close to that of copper. With small additions of alloys, its strength increases tremendously. Therefore, most acceptable products are actually alloyed titanium.

Table 16. **Titanium Alloys (Approximate properties)**

Type	Yield strength (psi)	Available forms
Ti-4Al-4Mn	133,000	wire, bar
Ti-5Al-2.5Sn	90,000 to 120,000	most forms
Ti-6Al-4V	100,000 to 130,000	most forms (extrusions)
Ti-7Al-4Mo	116,000 to 135,000	wire, bar
Ti-8Mn	110,000 to 140,000	sheets, strips

Ti — Titanium   Al — Aluminum   Mn — Manganese
Sn — Tin   V — Vanadium   Mo — Molybdenum

Titanium alloys are better known at present by the abbreviations of the important alloying elements and their percentages.

Weldability

Titanium is somewhat difficult to weld because of its undesirable ability, when heated, to absorb oxygen. Ordinary welding practices aren't suitable; therefore, the previously mentioned TIG and MIG arc-welding methods, which shield the hot metal from the atmosphere, must be employed.

Alloys

Alloy Ti-6Al-4V (a general-purpose alloy) was first used as the upper third of twelve-meter racing sailboat masts. This was done because of titanium's great ability to bend much farther under load than either steel or aluminum. (When a mast bends there is more control over the shape of the sail.) Titanium can bend up to four times as much as aluminum before breaking. Now, titanium is being used for entire masts and many products for all types of pleasure boats, not just sailboats. Although titanium products are for those who can afford them, advances in ease of production and in volume will hopefully bring the price down.

One development you may see in the near future is the introduction of nickel-titanium propellers, propeller shafts, hand tools, and cutlery.

ZINC

Zinc is considered the big protector of metals for two main reasons. In air, it corrodes at a very slow rate compared with that of iron and steel. This is due, again, to a thin oxide film that forms on its surface when it is exposed to the atmosphere. When galvanic corrosion takes place, zinc is the sacrificial metal (the anode) because of its position in the Galvanic Series.

Sacrificial Anodes

Zinc anodes cast in various forms (collars, plates, buttons, tear drops, and so forth) are made of zinc of a very high purity, with a small amount of aluminum added to improve castability and strength. Zincs are preferred over other anoditic materials because they corrode away only as fast as they must in protecting the metal to which they are attached. Zinc costs less and is more suited for cathodic protection on pleasure boats than magnesium, aluminum, or cadmium.

Zincs protect more than just the particular metal to which they are attached. For example, if a sacrificial zinc is attached to a steel rudder, the zinc also protects the copper alloy through-hull fitting in contact with the rudder shaft. The most important point to remember about zinc anodes is not to paint them if they are to work properly.

Coatings

There are several methods by which zinc is applied to steel and iron (mainly):

1. Electrolytically deposited (by electrogalvanizing or electroplating). This is usually a very thin coating.
2. By immersion in molten zinc (hot-dip galvanizing). This process is aided by the low melting point of zinc, which is about 787° F. Dipping produces the familiar mottled, angular, flaky appearance of such products. This process is recommended most for marine applications, and will provide protection for many years if not abused.
3. By spraying molten zinc (metallizing). The thickest coats can be produced by this process. This method is also recommended for protection of marine products.
4. By impregnation of zinc powder (sherardizing). The articles to be processed are buried in zinc powder, and with the help of heat, the lack of oxygen, and a little centrifugal force, they receive the zinc. This process forms a very thin covering and is not recommended for marine applications.
5. Zinc is also applied in the form of paints. Painting affords adequate protection if the paint is applied fairly thickly.

Note: What is important to long life is not the method of coating, but the thickness of the coating.

Note of Special Importance: Zinc-chrome-plated products (screws, nuts, bolts, and so on) just aren't sufficient on a boat. These are common products you see in little plastic bags in all the discount stores. The chrome has been added for a more shiny decorative appearance. The coating on these products is so thin and, because of the chrome, will galvanically corrode each other so fast that they are totally useless outdoors.

One situation where complications may set in is when galvanized-steel water or gas tanks are connected to copper lines. In an ocean environment, the copper can cause the zinc coating to corrode away faster than normal, leaving the tanks free to rust. This is even more true of any galvanized fitting that might be used in the lines, the reason being the corroding power of the large volume of copper over the smaller amount of zinc.

CADMIUM

Cadmium is another popular marine coating. It is comparable to zinc in many ways, but has several important differences:
1. It has a silverish-white appearance that is more attractive than zinc.
2. It has better resistance than zinc to salt-water splashing and marine atmospheric corrosion.
3. However, cadmium doesn't galvanically protect iron or steel in salt water as well as zinc.
4. It is applied generally by electroplating instead of dipping.
5. Cadmium is more expensive than zinc, but less expensive than other coating materials such as nickel or chromium.
6. Cadmium fumes and most cadmium compounds are poisonous; avoid breathing them when welding, and make sure that they have no contact with foods.

WIRE ROPE

1. Wire rope is always designated by the diameter of the rope.
2. Wire rope consists of either single wires or strands of wires wrapped concentrically around a single center wire or a strand of wires or a fiber (nonmetallic) core.
3. Wire rope with single wires is recommended for standing rigging (stays, shrouds, and so forth). The standard type is 1 x 19; for smaller sizes, 1 x 7 can be used. (Large sizes are very stiff.)

1 x 19   1 x 7   7 x 19   7 x 7

Table 17. Marine Wire Rope
(Approximate breaking strength of wire rope in lbs.)

SIZE	Galvanized improved plow-steel fiber core		Galvanized-steel aircraft cables			Stainless steel type 302 (type 304 has same strengths or slightly less)		
	6 x 7	6 x 19	1 x 19	7 x 7	7 x 19	1 x 7	1 x 19	7 x 7
1/32			180	110		150	150	110
3/64			370	270	270	370	330	270
1/16	350	350	500	480	480	500	500	480
5/64	480	480	800	650	650	800	800	650
3/32	750	800	1,200	900	920	1,200	1,200	900
7/64	980	1,050	1,600	1,250	1,250	1,600	1,600	1,250
1/8	1,150	1,250	2,100	1,700	2,000	2,100	2,100	1,700
5/32	1,900	2,000	3,300	2,500	2,800	3,300	3,300	2,400
3/16	2,700	2,900	4,700	3,700	4,100	4,700	4,700	3,700
7/32	3,750	4,000	6,300	4,800	5,600	6,300	6,300	4,900
1/4	4,800	5,100	8,200	6,100	7,000	8,500	8,200	6,200
9/32	6,100	6,700	10,300	7,400	8,000	10,600	10,300	7,600
5/16	7,500	7,800	12,500	9,200	9,800	13,200	12,500	8,800
3/8	10,500	11,100	18,000	13,300	14,300	18,000	17,500	11,700
7/16	14,000	15,000	23,400	16,000	17,500	26,000	23,400	15,500
1/2	18,500	19,000	31,000	20,500	22,700	33,700	29,800	21,200
9/16	23,400	24,000	38,500	25,500	28,500	42,000	36,200	26,500
5/8	28,000	30,000			35,000		47,000	32,500
3/4		43,000			49,500		67,500	46,000
1	70,000	75,000			85,200			76,500

Wire Rope

	Monel type 400			Stainless steel type 316		Aluminum bronze	Phosphor bronze	
7 x 19	1 x 19	7 x 7	7 x 19	1 x 19	7 x 7	1 x 19	7 x 7	1 x 19
				130			100	
270	200	135		300	240		200	110
480	330	220	220	450	360	360	300	170
650	510	340	350	740	500	500	420	230
900	750	480	520	1,090	700	800	670	370
1,250	1,000	650	680	1,450	975	1,020	850	450
1,700	1,350	850	870	1,900	1,300	1,340	1,100	650
2,400	2,100	1,300	1,350	3,000	2,000	2,050	1,700	1,000
3,700	3,000	1,900	1,950	4,250	2,850	2,900	2,450	1,450
4,900	4,100	2,600	2,650	5,700	3,750	3,800	3,000	1,950
6,400	5,300	3,400	3,500	7,600	4,800	4,800	3,900	2,500
7,800	6,900	4,300	4,400	9,500	5,900	5,900	4,900	3,150
9,000	8,300	5,300	5,400	11,500	7,500	7,300	6,100	3,850
12,000	12,000	7,650	7,800	15,600	10,800	10,300	8,600	5,500
16,300	16,400	10,400	10,700	21,000	14,300	14,000	11,700	7,400
22,800	21,500	13,600	13,900	26,500	19,500	17,800	15,400	9,600
28,500	27,000	17,200	17,600	32,000	24,400	22,400	18,700	11,800
35,000		20,600	21,800	42,000	30,000	27,400	23,000	14,500
48,500		29,600	31,400	60,000	42,500			20,600
85,200		52,500	55,800		70,500			35,500

4. Wire rope with strands of wires is quite flexible and is recommended for running rigging (halyards, sheets, and so on). Cables, davit-winch cables, and mooring lines are other possible uses. The most common number of wires in a strand are 7 and 19, known respectively as 7 x 7 and 7 x 19 wire rope. For special purposes, there is also available wire rope with 12, 24, 37, and 42, wires per strand (6 x 12, 6 x 24, 6 x 37, and 6 x 42). The 6 x 42 construction is known as tiller rope, but because of its low strength in the smaller sizes it isn't used much in pleasure boats. The 7 x 19 construction is best suited for steering cables.
5. Some types of wire rope have fiber cores. If the first digit is a 6 — 6 x 19, for example — the center may be of a soft fiber material. Fiber core rope is usually designated by the capital letters *FC* (Examples: 6 x 19 FC, or 6 x 7 FC.)
6. Wire ropes are lubricated during manufacture to minimize internal friction. Regular applications of oil will help extend their life.
7. An important feature of wire rope, that has now become an industry standard, is preforming. Each strand or wire is bent to form the approximate curving position it will assume in the finished rope before it is put together. Preformed wire rope should not fray when you cut it, and lies straighter than non-preformed wire rope. Preformed wire rope is usually marked as such.
8. Plastic-coated wire rope is advisable for several applications. One standard use is in steering cables. Sailboat builders or manufacturers might consider using plastic-covered cables for rigging: It not only increases the corrosion resistance of the cables, but it can also help prevent sails from chafing.
9. Table 17 lists the approximate breaking strengths of various materials, types, and constructions of wire ropes that are used or may be used for boating applications. It may be hard to locate some of the more exotic types and sizes listed, mainly because of the

lack of demand for them. There are numerous other types, with quite a diversity of constructions, besides the ones listed here. However, these other types have special characteristics that make them more suitable for such uses as elevator cables. Note that the strengths given in the table are only approximate and may vary depending on the manufacturer.

CHAIN

There are two forms of chain, stud-link chain and open-link chain. Even though stud-link chain is known as anchor chain, most pleasure boatmen never have occasion to use it because it's for the "big boys" (ocean liners and vessels of comparable size). There are many types and brands of open-link chain. Most (generally in a galvanized condition) are used for mooring and anchoring.

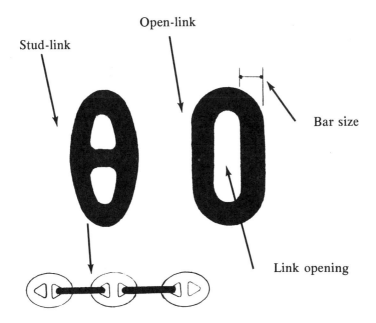

Table 18. Strength of Various Sizes of Chain
(Chain-breaking strains depend on size and grade of chain, type of steel, type of chain, and manufacturer.)

Size	Minimum breaking strain could be as high as or higher than	Minimum breaking strain could be as low or lower than
3/16	3,200	2,000
1/4	10,000	3,600
5/16	11,500	5,000
3/8	19,000	7,200
7/16	20,500	8,400
1/2	32,000	13,000
5/8	50,000	20,000
3/4	64,000	28,000

Chain size is referred to by its bar diameter. As chain sizes increase, the link openings (the length and width of the center hole of the link) increase proportionally. The link openings also vary according to type and grade. For example, long-link chain and (as the name implies) buoy chain have longer links than average. Familiar grades of steel chain are high-test, BBB, and proof-coil, the latter being the weakest.

When purchasing chain, don't equate a specific chain size with one standard strength. For high-test steel chain, the recommended minimum breaking strain for the 3/8" size, the strain is 9,800 pounds. There are many considerations that point out that the person buying chain for mooring and anchoring should inquire about the exact working load and breaking strain of the particular size of chain in question. That this is necessary should be obvious from the data in Table 18, the numerous grades and brand names prevalent in the industry, and the use of the word "recommended" by suppliers. If the dealer can't show you the manufacturer's limit, don't buy the product.

Another important point to remember is that the recommended working load of chain is much less than the breaking strain, and the endurance limit is even less. Working loads are only about 20 percent to 30 percent of the breaking strain. A grade of chain equivalent to or higher than BBB steel chain is recommended for anchoring and mooring.

PURCHASING INFORMATION

If you can't find a supplier for the particular metal or metal product you want, go to the library and look for the names of suppliers in the *Thomas Register* or *Macrae's Blue Book*. These are corporate indexes with addresses, products, trade names, and some manufacturer's catalogs. If a particular company doesn't have an office in your area, write to them and they will tell you of a distributor near you.

Besides a marine hardware store or metal-supply house, a very excellent and reasonable source of metals is your local scrapyard. The yard most likely to have what you are looking for is one that advertises that it will buy nonferrous metal or metal products such as radiator cores, old batteries, or scrap brass, bronze, aluminum, and so on. Even better is a yard that has junk airplanes or government surplus equipment. If you haven't had this experience, you will be amazed at what you can find and make from scrap materials.

Table 19. Chain (Approximate properties)

Size (inches)	Glavanized High-test steel chain		Galvanized BBB steel chain		Galvanized Proof-coil steel chain	
	Work load	Break load	Work load	Break load	Work load	Break load
1/4	2,500	7,700	1,300	5,300	1,150	4,700
5/16	4,000	11,500	1,950	7,800	1,700	7,000
3/8	5,000	16,200	2,750	11,000	2,400	9,800
7/16	6,500	20,500	3,600	14,500	3,200	13,000
1/2	8,000	26,000	4,750	19,000	4,200	17,000
5/8	11,500	36,900	7,200	29,000	6,350	25,500
3/4	16,000	50,400	10,200	41,000	9,100	26,500

MAGNETS

Often, a boat owner or builder wishing to sell his boat will cover up signs of rust, mistakes, cheap materials, and corrosion, with lots of paint. So a good tool to have along when you are buying a boat or metal product is a magnet. Most noncorrosive metals, such as aluminum, bronze, copper, and the 300 series of stainless steels, are nonmagnetic. If your magnet is attracted to a bright and shining object, you can be quite sure the object may rust before too long. (Apply this principle to all but hot-dipped galvanized parts.) *Caution:* Keep the magnet away from the boat's compass.

Also, suppose you are buying a fishing rod and reel. A magnet can tell you if there are any, or many ferrous parts (iron or steel) that will corrode or rust. In addition, determine how many different metals were used. This might indicate its corrosion resistance. (Dissimilar metals form a galvanic couple, and corrode.)

GLOSSARY OF TERMS

Alloy. A substance composed of two or more metals (a base metal and a smaller amount of another metal, or in some cases, the addition of minor elements), the combination of which yields durability, strength, or some other desired quality.

Annealing. Annealing is a form of heat treatment. The purpose of annealing is to put a metal in its soft condition. There are different annealing processes for different metals. Broadly speaking, the metals are heated to high temperatures and are subsequently cooled to soften them and make them less brittle. The reason this process is required is that formed, forged, or machined metals have accumulated stresses that must be relieved.

Annealing differs from tempering in that it is more of an intermediate process to improve further the metal's machinability, rollability, or cold-working qualities, although metals are often put to use in their annealed condition.

Tempering is aimed more at the final hardness and strength required of the metal.

Artificial Aging (See also *Natural Aging*). Aging at high temperatures to improve strength and other characteristics of a particular metal.

Burnishing. Burnishing is a mechanical polishing technique. Very smooth steel balls or specially shaped smooth tools that are applied with pressure create friction and heat that leaves the metal surface smoothly compressed and somewhat brightened. Burnished bronze is an example of a metal so treated.

Case Hardening. This term means just what the words say: to harden the thin outer case or shell of a piece of metal. This usually refers to steel products such as nuts, bolts, and fasteners. Carbon, the most important element in determining the ultimate hardness of steel, is added to the outer surface of low carbon or mild steel. This outer surface is then heat-treated so that it hardens and forms a case around the softer core.

Cavitation Erosion Corrosion. This is a combination of mechanical (wear and erosion) and extremely intensified corrosion damage due to high-velocity sea water. Propeller blades and shafts — and to a much lesser extent, struts and rudders — are typical metal boat parts subject to cavitation-erosion-corrosion. Alloys that have good resistance to this type of corrosion are 316 stainless steel, titanium, nickel-aluminum bronze, aluminum bronze, manganese-nickel-aluminum bronze, nickel copper, and copper-nickel alloys.

Cold Worked. A mechanical process for tempering metals, usually by running the metal through a series of rollers.

Commercial Sizes. These are preferred manufacturing items, and orders for their production are accepted without special inquiry. The manufacturer permanently keeps the tools (dies, jigs, molds, etc.) for ready production of these sizes. He doesn't stock the parts because of the low or infrequent demand. ("Parts," in this case, refers to such items as nuts, bolts, turnbuckles, wire rope, etc., that are made in numerous sizes).

Drawn Tube. A tube brought to final dimensions by drawing it through a die.

Ductility. The ability of a material to withstand plastic deformation (such as stretching, bending, flexing, twisting, denting, and crushing) without rupture.

Elastic Modulus (Modulus of Elasticity). The measure of a material's ability to stretch without deforming permanently.

Electropolishing. This is an electrolytic process that removes metal irregularities from the surface of metals, leaving them shiny and bright. Electropolishing is considered better than mechanical methods of polishing because it leaves the metal free of scratches and improves corrosion resistance. Most of the alloys mentioned in this text can be electropolished.

Extrusion. A process whereby a product is formed by forcing metal through an orifice in a die.

Fatigue. The failure of metal due to repeated alternating stress. Many failures of marine metals in use are caused by fatigue aided by corrosion.

Heat Treatment. A process of heating and cooling metals in order to obtain certain desired properties, such as hardness, toughness, and workability

Investment Casting. This is a very accurate casting process used to make such products as shackles, which have intricate curves that are difficult to machine. First, a pattern is made of the desired object. Next, a permanent mold is made in which wax or a similar substance can be used to make duplicate copies. These sacrificial copies (the investment) are embedded in sand, then melted away to form the cavity in which the finished product is cast. Investment casting can be of either ferrous or nonferrous alloys.

Machinability. Machinability refers to the ease with which metal can be removed in such operations as drilling, threading, reaming, sawing, and turning, resulting in a good finish. Good machinability is usually due to the addition of a small amount of some element such as lead or silicon to the base metal.

Malleability. A metal is malleable if it is capable of being shaped or extended by beating with a hammer or by the pressure of rollers, without breaking or cracking.

Manufacturing Limits. The maximum normal range of alloys, tempers, and sizes that a particular manufacturer is equipped to produce.

Natural Aging. This is a hardening and strengthening phenomenon associated with the process of precipitation hardening. Such metals as aluminum and stainless steel can be produced with overadditions (so to speak) of alloying elements that chemically strengthen a metal over a period of time. In natural aging, a metal is left to sit at room temperature until it hardens (precipitates). This process can be hastened by heat and is called artificial aging.

Quenching. Rapid cooling, usually by immersion in water.

Standard Sizes. These are items priced for large-quantity production, and are stocked in warehouses.

Strain Hardening (as used in this text). An increase in hardness and strength as a result of stresses of working and forming operations on aluminum and its alloys.

Tempering. A term used to describe several methods of bringing metals to their desired texture, degree of strength, degree of toughness, degree of hardness, and consistency. Tempering methods mentioned in this guide are work hardening (cold-worked), direct heat treatment, and precipitation hardening. Tempering is not to be confused with annealing, which leaves a metal in its soft condition (see *Annealing*).

Tensile Strength. This is the minimum amount of resistance offered by a particular metal (or material) to the tensile stress (pulling force) required to break it. Presently, in the United States, the standard way of designating tensile strength is in pounds per square inch (psi) of cross-sectional area. The adoption of the metric system may soon change this.

Weldability. A question that must be asked before proceeding to weld a particular metal is, "Is it weldable?" Will the heating and cooling affect the strength of the metal or the weld deposit? How difficult will it be to achieve an adequate weld?

The difference between brazing, soldering, and welding:

> *Brazing.* The two metals to be joined are not melted, but a filler metal that melts at a temperature below the melting point of the metals to be joined is used. This metal is drawn into the

close-fitting joint by capillary action. The filler metal melts above 800° F. Brazing is not quite as strong as welding, but many brazed joints often approach the strength of a welded joint.

Soldering. Soldering is the joining of two metals with a filler metal that melts below 800° F.

Welding. The two metals to be joined are heated to their melting points and are fused with a filler metal or, in some cases, pressed together without the use of a filler metal.

MIG Welding (*metal-inert-gas arc welding*). A rod or wire produces the arc and filler metal. A torch provides either argon, helium, or carbon dioxide gas, which acts as a shield to protect the molten metal from oxygen contamination.

TIG Welding (*tungsten-inert-gas arc welding*). A tungsten electrode provides the arc, but is not consumed. The filler metal comes from a wire or rod. A torch provides either argon, helium, or carbon dioxide gas, which flows around the electrode and acts as a shield to protect the molten metal from oxygen that can oxidize some of the alloying elements and leave the weld more susceptible to corrosion. Oxygen can also cause embrittlement of some metals.

Shielded Metal-Arc Welding. An arc is struck between a current-carrying electrode and the workpiece, or workpieces. The arc is sufficiently hot to melt the tip of the long, pencil-shaped electrode and a small area of the workpiece. The electrode is slowly consumed as it provides a filler metal to complete the bond. The electrode is usually a similar metal to that being welded. A flux is responsible for shielding the weld from atmospheric contamination, cleaning, and metallurgical control. The flux is in the form of a coating on the electrode.

Work Hardening. If a metal or alloy is rolled, pressed, pulled, or hammered beyond its yield point, it usually becomes harder and stronger, but its ductility is reduced.

Wrought. Wrought metals are shaped or shapable by hammering, beating, or rolling. Wrought metal includes sheet foil, plate, extrusions, tubes, forgings, rods, bars, and wire. Castings are not wrought metal. Wrought aluminum, iron, and steel are examples of wrought metal.

Yield Strength (*Yield Point*). This is the minimum unit stress (tested as a pulling force) at which a structural material (metal) will deform without a further increase in its load. (The amount of strength required to stretch the metal's length by 2 percent is the standard test.) This amount of force is called the yield point of the metal. Some materials do not have this point; they merely break. For the ones that do, it is a well-defined value.

Index

Active state: corrosion of stainless steel, 31-32
Admiralty brass: the galvanic series, 2; composition and usage, 18, 21
Admiralty bronze; composition, 18
Alloy Casting Institute (ACI): identification system, 35
Aluminum: effects of galvanic corrosion, 3; corrosion resistance, 3, 6-7, 8; structural use, 6; increasing strength of, 7-8; non-heat treatable alloys, 7-14; post-weld heat treatment, 8; bottom painting, 11; tempering of alloys, 11-15
Aluminum alloys: description, uses, strength, 10, 13
Aluminum alloy 6061: uses and description, 8, 9, 13
Aluminum bronze: qualities of, 22
Aluminum-silicon bronze: qualities of, 22
American Iron and Steel Institute (AISI): identification system, 30
Anodes: magnesium, 24; zinc (sacrificial), 45
Anodic: place and purpose in galvanic series, 3; relationship to pitting, 4-5
Anodic finishes of aluminum: anodizing, 9
Antifouling paints: protection of aluminum, 11; composition, 17
Atmospheric corrosion: contributors, 6; weathering of aluminum, 8; *See also* Oxide films.

Bolts and nuts: brass, 18; Tobin bronze, 23; tin bronze, 25; monel, 26; stainless steel, 28-29; zinc coated, 46; case hardening, 55; sizes, 55. *See also* Silicon bronze.
Brazing: explanation, 57-58
Burnishing: usage, 55
Butyl rubber: use of, 9. *See also* Separators.

Carbon dioxide: corrosion above waterline, 6
Case hardening: nuts, bolts, 55
Cathodic: place and purpose in galvanic series, 3; relationship to pitting, 4-5
Checking metals on new and used boats: for proper use, 19
Chicken wire: poultry netting, 40
Chloride ions: their part in corrosion, 32
Chlorine: corrosion above waterline, 6
Cleats: aluminum, 16
Coating of metal: gold, 19; corrosion protection, 24; with zinc, 45-46; with cadmium, 47
Coloring aluminum: anodizing, 9; painting, 11
Composition of brass and bronze, 17-18
Contact corrosion: galvanic corrosion of aluminum, 8-9. *See also* Separators.
Copper lines and tubes: corrosion of, 46. *See also* Cupronickel.
Copper powder: use in antifouling paints, 17
Corporate indexes: names, use of, 53
Corrosion: basic definition, 1
Cracked or crackled metal: cause and speed of corrosion, 5; in aluminum, 8. *See also* Crevice corrosion and Stress corrosion.
Crevice corrosion: stainless steel, 4, 31; aluminum, 8

Cupronickel: its place in galvanic series, 2; content, 17, 21; as a substitute for copper, 23

Decomposition of metals: cause, 1; copper (uses to avoid), 17; brass, 18
Decorative extras: imitation gold, 19. *See also* Coating of metals.

Electrogalvanizing (zinc), 45-46
Electrolysis: electrochemical relationship, 1
Electrolyte: relationship to corrosion, 1
Electrolytic-tough-pitch: copper, 16, 21
Electroplating: zinc, 45-46; cadmium, 49
Electropolishing procedure, 56
Etching of aluminum, 11

Ferric Oxide: corrosion of iron and steel, 6
Ferrocement compared with fiberglass, 42-43
Fiber cores: of wire rope, 50
Fishing rods: 302 stainless steel, 28; purchasing information, 54

Galvanic corrosion: magnesium, 24; between stainless steel and aluminum, 32; titanium, 43; cadmium, 47. *See also* Contact corrosion.

Halyards: wire rope, 50
Hand railings: aluminum, 13
Hard-coat anodized: description, 9
Hardness: designations for aluminum, 12
Heat-treatable alloys: aluminum, 7-14
Heat treatment: stainless steel, 30. *See also* Welding of stainless steel (sensitization).

High-zinc brass: types, 18
Holes in metals due to corrosion (dezincification), 5. *See also* Decomposition of metals, and Pitting.
Hot-dip galvanizing: steel and iron, 46
Hull material: aluminum, 6-7; steel in conjunction with cement, 35-36

Identification systems of metals (digits and letters): aluminum, 11-15; cast aluminum, 15-16; copper alloys, 17; magnesium, 24; stainless steel, 30; precipitation-hardening stainless steel, 33; cast stainless steel, 34; titanium, 44
Improving corrosion resistance: anodizing of aluminum, 9; of copper, 16-17; phosphor bronze, 22; chrome plating, 23; of stainless steel, 31; of zinc, 45
Investment castings: description, 56

Least noble metals: place and purpose in galvanic series, 3

Machinability of brass: addition of lead, 19
Magnesium: uses, 2, 3, 24, 26
Magnets: metal detectors, 54
Manganese bronze: galvanic series, 2; composition and usage, 17, 18, 21
Marine metals: reasons for their importance, 1
Marine microorganisms: poisoning (antifouling), 17
Metallizing: iron and steel, 46
Modulus of elasticity: explanation, 56
Monel: its place in galvanic series, 2; general discussion, 26-27; corro-

sion above waterline, 26; nails, 26; wire rope, 48-49

Naval brass: its place in the galvanic series, 2; composition and usage, 17, 21
Neoprene tapes; use of, 9
Noble metals: place and purpose in galvanic series, 3

Open link chain, 51.
Oxide films: and passivity, 3-4; breakdown of (as an example of crevice corrosion), 4; on aluminum, 8-9; on stainless steel, 31; on titanium, 43. *See also* Anodic finishes.

Painting: of aluminum, 11; with copper powder, 17; zinc, 45; with zinc, 46; and magnets, 54
Passivated stainless steel; passive state of, 31
Passivity: required oxygen concentration, 3-4. *See also* Oxide films.
Phosphor bronze: properties of, 19-22
Phosphorous-deoxidized copper: composition and usage, 16, 21
Pitting: cause, 4-5; gasoline tanks and lines (cause), 4-5; aluminum, 8; Monel, 26; stainless steel, 32
Plastic-covered cables: wire rope, 50
Plating of metal: gold, 19; chrome, 23
Porthole frames: Admiralty brass, 21
Precipitation-hardening stainless steel: types and uses, 29, 33-34
Preformed wire rope, 50
Propellers: bronze, 25; Monel, 26; nickel titanium, 45
Propeller blades: cavitation erosion corrosion, 55
Purchasing information: brass, 18; reinforcing wire, 43; chain,

52-53; metal and metal products, 53

Reinforcing steels: ferrocement, 36-42
Rudder quadrant: manganese bronze, see note 1, page 21. *See also* Dezincification.
Rusting: of iron and steel, 6

Sailboat masts: aluminum, 9, 10, 14; titanium, 44-45
Scaling: dezincification of metals, 5
Scrapyards (junkyards): metals source, 53
Screws: brass, 18; copper alloys, 20, 21, see note 1, page 21; Monel, 26; stainless steel, 28-29; zinc coated, 46. *See also* Bolts and nuts.
Selecting metals, 54
Sensitization: a problem when welding stainless steel, 33
Separators: for prevention of galvanic corrosion between aluminum and stainless steel, 3, 8-9
Shackles: from investment castings, 56
Shafts (propeller, rudder, etc.): copper alloys, 20-21; Tobin bronze, 23; Monel, 26; stainless steels, 28-29, 32; nickel-titanium, 45; corrosion of, 55
Sheets: wire rope, 50
Sherardizing iron and steel, 46
Shielded metal-arc welding procedure, 58
Shrouds: wire, 50
Silicon bronze: its place in galvanic series, 2; composition and usage, (nuts, bolts, and nails), 20; qualities of, 22
Soldering: explanation, 58
Solution heat-treatment of aluminum: temperature, 7-8; designations, 14

Spars: aluminum, 9, 10, 14; titanium, 44-45
Stainless steel: effects of galvanic corrosion, 3, 32; classes of, 27, 30, 31; appearance of, 32.
Stainless steel alloy 18-8: composition, 30
Stainless steel alloy 302: 28, 30-31.
Stainless steel alloy 316: 28, 30-31, 32
Stanchions: aluminum, 13
Standing rigging: wire rope, 47, 50
Stays: wire, 50
Steel rudders: protection of, 45
Strength of castings: aluminum, 16; copper alloys, 25; Monel, 26-27; stainless steel, 35; zinc, 45
Stress and corrosion of metals, 4. *See also* Cracked or crackled metal.
Stud-link chain: description, 51
Sulphur: corrosion above the waterline, 6

Tempering: general procedure, 57
Tensile strength: explanation, 57
Through-hull fittings: problems, 6; dezincification, 6
Tiller rope: wire, 50
Titanium: its place in galvanic series, 2; effects of galvanic corrosion, 3; description and use of alloys, 43-44
Tobin bronze: uses of, 23
Tubing: copper, 16-17
Tufnol washers: use of, 9
Turnbuckles: copper alloys, 20-21, 23; stainless steel, 28-29

Visual identification of metals: brass, 18-19; cupronickel, 18-19; cadmium, 47
Water tanks: cartridge brass, 21
Weathering: improving corrosion resistance, 9
Welded wire fabric: reinforcing, 39.
Welding: of aluminum, 7-8; of brass, 19; of phosphor bronze, 22; of magnesium, 26; of Monel, 26; of stainless steel, 30, 33; of titanium, 44; of cadmium, 47; procedures and descriptions, 57-58
    MIG welding: uses, 7, 33, 44; procedure, 58
    TIG welding; uses, 7, 22, 26, 33, 44; procedure, 58
Welding rods: stainless steel, 33
Wiggle rod: lower strength re-bars, 38
Winch cables: wire rope, 50
Winches: aluminum, 16
Wire cloth: sizes and styles, 41
Wire rope: aluminum bronze, Monel, phosphor bronze, stainless steel, 48-49; extending the life of, 50
Wrought: description, 59

Yield strength: explanation, 59

Zinc: uses, 2, 3, 10, 15, 17, 18, 24, 45-46; effects of galvanic corrosion, 3; corrosion problems, 5; strengthening castings, 45
Zinc collars: anodes, 45
Zinc-chrome plated iron and steel: qualities of, 46. *See also* Bolts and nuts.